NINJA SHURIKEN THROWING

THE
WEAPON
OF
STEALTH

NINJA SHURIKEN THROWING

SID CAMPBELL

PALADIN PRESS
BOULDER, COLORADO

Ninja Shuriken Throwing
The Weapon of Stealth
by Sid Campbell
Copyright © 1984 by Sid Campbell

ISBN 0-87364-273-2
Printed in the United States of America

Published by Paladin Press, a division of
Paladin Enterprises, Inc., P.O. Box 1307,
Boulder, Colorado 80306, USA.
(303) 443-7250

Direct inquiries and/or orders to the above address.

All rights reserved. Except for use in a review, no
portion of this book may be reproduced in any form
without the express written permission of the publisher.

Neither the author nor the publisher assumes
any responsibility for the use or misuse of
information contained in this book.

Photographs by Ed Reynolds III and Ed Evans

Dedication

This book is dedicated to the martial artists who seek to broaden their knowledge, wisdom, and exceptional skills beyond what is expected of them.

Acknowledgment

I would like to express my sincere gratitude to my dedicated students who have participated in the many practice sessions and who gave so willingly of their time and effort in creating the photographic concepts, artistic renderings, and ideas presented here.

I am happy to have been able to share with them much of the martial arts knowledge I have acquired.

Contents

    Preface   ix
    Introduction   1
1. Brief History of Ninjitsu   3
2. The Ninja   7
3. What Is a Shuriken?   11
4. Shuriken Composition   15
5. Gripping the Shuriken   19
6. Throwing Positions   35
7. Drawing the Shuriken   53
8. Trajectory Methods   57
9. Penetration Characteristics   73
10. Sighting Methods   79
11. Accuracy   83
12. Concentration   91
13. Spike Shuriken Trajectory   95
14. Fighting Tactics   101
15. Target Areas   117
16. Shuriken Concealment   123
17. Moving Targets   125
18. Rapid Fire   129
19. Multiple-Shuriken Throwing Techniques   131
20. Target Competition   133
21. Additional Shuriken Considerations   137
22. Conclusion   139

Warning

Possession and/or concealment of a shuriken is considered a criminal offense in certain states and localities.

Preface

Having spent over twenty years in the study and search of martial arts knowledge, I realize that it would be impossible, even in a period of many lifetimes, to become knowledgeable and skillful in every known aspect of every martial arts method and style. I have had the pleasure of sharing information with many fine martial artists throughout the past twenty years, and I'm sure to meet many more in the years to come.

This book will explore the many discoveries I have made concerning the "Throwing Stars" or shuriken that were used so expertly by the Ninja of the past. Over the years, I have been fascinated with the shuriken, probably because when used in the proper manner, it has the capability to "stick" every time it is thrown. Like the boomerang, it returns to the thrower when properly thrown.

At first it was with some reservation that I could seriously consider this seemingly innocent star a "real" weapon, as it appeared to be more of a toy. Many of the mass-produced shuriken today are still somewhat toylike because of the manner in which they are produced, and it is doubtful they could be used effectively in the ways I will describe in this book. To discover many of the truly effective methods of using the shuriken, I had metal workers and skilled machinists hand-make the weapons in a manner that required many hours of painstaking labor. Once the work was completed, I conducted an in-depth study of the balance, applications, accuracy, and techniques of the handmade shuriken.

The techniques and methods of using the shuriken presented in this text will undoubtedly vary from person to person. It is quite possible that no new revolutionary techniques will be presented that were not utilized by a highly trained Ninja, but I have tried to present a factual, practical, and realistic approach to using the shuriken today. Many of the technical aspects relating to the trajectory, velocity, and tactical elements of this unique weapon have been presented in order to provide insight into the concepts involved in the use of the shuriken. One's skill with this weapon will be achieved if conscious effort and practice are maintained on a regular basis.

I encourage each and every serious shuriken artist to develop patience and diligence toward his training and use this book as a guide toward increasing both physical and mental skills while seeking perfection with the shuriken.

SID CAMPBELL

Introduction

I have chosen to concentrate on only one single weapon of the Ninja's arsenal with hopes of making the reader aware of the vast amount of military and martial arts skills that the elite Ninja possessed.

The Ninja were extremely skilled individuals who totally dedicated themselves to the requirements of their clients. In all probability, the shuriken played a very important role in their weapon arsenal because such a weapon was silent, deadly, and could be easily concealed. It is virtually certain that these highly skilled "Shadow Warriors" spent the majority of their spare time sharpening and maintaining their weapon skills in the event they were called on to supply a particular need.

The tactics, methods, and technical skills presented in this book will give the serious weapon practitioner a chance to learn of the skills that were undoubtedly possessed by the versatile Ninja of feudal Japan and to enlighten the reader as to the sincerity which is required to master shuriken throwing.

Prepare to develop these unique weapon skills, which could possibly save your life in the event it becomes necessary to utilize a simple, silent, and effective self-defense weapon.

Many of the techniques and concepts presented here can also be used in a contest of skills with fellow martial artists in much the same way that darts can be used in games of skill. This method of practice requires very little space and would give the practitioners a chance to maintain maximum proficiency in such skills as timing, eye/hand coordination, accuracy, power, speed, and a host of other physical attributes required for empty-hand styles of martial arts.

1. Brief History of Ninjitsu

Most noted authorities believe that the mystical art of Ninjitsu originated over two thousand years ago, yet no one is sure of the exact date.

There is reference to such an art in the *Ping Fa (Art of War)*, an ancient Chinese book which details the military sciences of spying and its unique methods. The *Ping Fa* was written by Sun Tzu, who lived between 500 and 300 B.C.

The art of Ninjitsu is thought to have been introduced to Japan in the sixth century. At this time, Prince Shotoku (593 to 622 A.D.) employed agents skilled in the art of spying to secretly report activities relating to civil disputes. Many of these disputes concerned land ownership, doctrines, or positions of power. It is probably due to the *Yamabushi* ("mountain warriors") that the art of Ninjitsu was preserved and developed into a complete and detailed science of spying.

In martial arts circles, the art of Ninjitsu is more commonly referred to as the "Art of Stealth," and many people associate this art with the methods employed by the assassins of feudal Japan. Every conceivable method of assassination, spying, breaking and entering, sabotage, stealth methods, concealment methods, weapon applications, and warfare tactics were employed by the Ninjitsu practitioners.

Between 1192 and 1590 A.D., Japan experienced four centuries of civil strife and the many warlords of Japan employed the services of the Ninja (Ninjitsu practitioner) to secure and maintain their positions of power. Even though there were many clans or schools of Ninjitsu throughout the centuries, only a few are responsible for bringing the Ninjitsu art of stealth and espionage into the twentieth century. As with many of the arts of Japan, certain dedicated individuals or schools propagated and preserved the traditions of their ancestors, and the art of Ninjitsu is certainly no exception.

The two primary factions responsible for the majority of the Ninjitsu arts were the Iga and Koga clans. These clans were in essence provinces or towns located on Honshu, the largest island of Japan. The province of Iga is today known as the Mie province and is located on the southern end of Honshu; the Koga clan or province is located near Tokyo.

Since the Iga and Koga were the largest Ninjitsu clans, it is natural to assume that their arts have survived into the twentieth century due to their strength and considerable size.

In addition to the shuriken, weapons utilized by the Ninja included the sword, bow and arrow, spear, broadbladed spear, staff, fast-sword drawing methods, chain and scythe. Many unarmed combat methods were also practiced.

Not only were the Ninja masters of the traditional weapon arts, but they were also exceptionally skilled in special-purpose weapons, such as daggers, darts, brass knuckles, dirks, shuriken of many designs, garrotes, lead-weighted bamboo staves, caltrops, grappling hooks, guns, grenades, smoke bombs, eye-blinding powders, acid guns constructed of bamboo, poisons, and other secret devices of assassination.

To become skillful in the art of Ninjitsu, a warrior had to be skilled in the art of camouflage and disguise; accomplish incredible physical feats, such as scaling high castle walls; acquire expertise in mountain climbing and running long distances; remain submerged under water for several minutes at a time; make weapons from makeshift resources; and become an exceptional escape artist in the event of capture.

Most of the Ninjitsu clans ran their organizations in much the same manner as an army, and the training was rigorous and extremely demanding. The Ninjitsu artist had to dedicate his life to his profession and be willing to die for the sake of completing a mission for the warlord or party employing his services. In the event of capture, many Ninja warriors would use concealed explosives to take their lives and destroy their facial features so their identities or clans would not be divulged to the enemy.

The art of Ninjitsu is enjoying a resurgence in popularity throughout a large part of the West since the end of World War II, partly due to the fact that many servicemen stationed in Japan after the war were exposed to the martial arts.

2. The Ninja

The epitome of stealth and espionage, the Ninja was attired either in black or any suitable color that matched his surroundings.

The mystique of the Ninja has never in all probability been paralleled, even today, by the most skilled military specialist. Many of the skilled military men of today are trained in specific areas of expertise, whereas the Ninja was a master of virtually every form of military warfare.

In addition to their military skills, they were strongly committed to their missions, enlightened in the ways of nature, and they upheld the principles of their convictions through an established code of ethics.

From an early age, they received a thorough practical education and were well versed in the traditional cultural arts of the times. These arts included painting, tea ceremonies, flower arranging, playing musical instruments, performing traditional Japanese dance, and telling stories and jokes. This traditional cultural education also made it possible for the Ninja to assume numerous identities in their training for possible espionage assignments.

The organizational structure was generally considered to be a familial one, and the novice Ninja usually began his initial training as early as five years old. From his early years

The Ninja could adapt to any situation and was prepared for anything at all times.

of training, he was committed to a lifelong profession that would require virtually every waking hour in the development and perfection of the extraordinary skills of his trade. Every facet of military warfare was taught to the Ninja, and each "art" was considered to be a science within itself. Each of these sciences had to be thoroughly understood and perfected before it could be mastered.

Since this book deals with only one of the weapons utilized by the infamous Ninja, it will not be the purpose of this section to give a detailed description of every facet of training, tactical application, or methods of operation utilized by the Ninja.

The Ninja were known by many names, such as Shadow Warriors, *Shinobi, Kunoichi* (female Ninja), Shadows, Iga Ninja, and Koga Ninja. Regardless of the title bestowed on them, the Ninja were feared by warlords, shoguns, and *Daimyo* (feudal lord) alike.

Folk tales describe the Ninja as capable of flying, walking on water, disappearing through solid rock walls, living underwater, and even vanishing in a puff of smoke. As outlandish as these tales may seem, there were many logical explanations to substantiate them. With the use of explosives, powders, specialized flotation devices, and underwater air bags the Ninja devised, it would appear to the unknowledgeable person that these incredible feats had actually been accomplished.

It should be mentioned that the Ninja were in essence professional hired assassins who based their existence on the services they could perform for the benefit of the Daimyo, warlord, or shogun willing to finance their services.

In the modern world, the Ninja and their organizational structure parallel many of the military services existing in the world today except that the services of the Ninja were available to the highest bidder. The Ninja, however, was more skilled in every facet of the military arts than the majority of personnel in the armed services today.

3. What Is A Shuriken?

The word shuriken is derived from the Japanese and is defined as a sharp-pointed or blade-throwing instrument. Although there are many shapes and designs utilized in shuriken-jutsu, there are essentially two categories of this weapon: the star-shaped design and the spike-shaped design.

Until recently, there were ten distinct star-shaped styles favored by the Ninja; at present there are hundreds of shuriken designs. Some of the early shuriken were cross-shaped, four-pointed, six-pointed, eight-pointed, and ten-pointed, while others were triangular, swastika-shaped, hexagonal, pentagonal, and three-pointed.

The shapes were for the most part determined by the particular needs and objectives of the Ninja warriors who designed them.

Some of the star-shaped shuriken had a round or square hole through the middle of the weapon which served multiple purposes. First, it permitted the Ninja to carry numerous shuriken on a leather thong or spiked object so that they could be readily dispensed if the need arose. Since the shuriken were mounted in an orderly and accessible fashion, they could be dispensed in a rapid-fire fashion.

Second, the holes permitted for more utilitarian purposes, such as removing nails or other securing pins of a lock or other security device. This came in handy when a Ninja tried to gain access to a building as part of an assigned mission.

Of all the star-shaped shuriken, the ones that seem to be the most effective are the ones having four- or five-point designs, since these make deeper penetration possible.

The spiked shuriken are a bit more simple in design, yet they are more difficult to master because the number of points available for sticking is limited. Some of these shuriken were pointed on one end, while others were designed with both ends pointed. The advantage of the spiked form, however, is its ability to penetrate far deeper than the star shape, thereby being more lethal when used by a skilled shuriken-jutsu artist.

The spiked forms were undoubtedly used in the same manner as the caltrops (small, multi-pointed devices that were placed in the enemy's path to injure and slow him down). Since it is more difficult to master this type of shuriken, a thorough understanding of trajectory methods, distance from target, and balance characteristics must be developed.

The range for both the star and spike shuriken is about thirty feet, though this distance will vary somewhat from person to person. The primary factor, in any case, is to exercise accuracy, and the thirty-foot range is generally the maximum effective range in which the skilled practitioner can expect to achieve hits on a consistent level.

In earlier times, the Ninja usually carried nine shuriken due to the fact that nine was considered to be a lucky number and the Ninja had to be prepared in case he encountered more than one enemy. The majority of shuriken designed for warfare were primarily used to injure rather than kill the enemy, unless the points were long and sharp and the thrower was extremely accurate in his throwing technique.

Penetration would have to be deep and directed to such vulnerable areas as the throat, heart, or main artery before

The Ninja is prepared to throw his shuriken in case of danger.

death would result from a shuriken wound. Even though the shuriken could be lethal, it was primarily effective in maiming or stopping a pursuer long enough for a Ninja to escape.

4. Shuriken Composition

Using their unique ingenuity, it was possible for the Ninja to create shuriken from virtually every type or configuration of metal that existed in their times. Metals such as bronze, brass, iron, and steel were the most common substances from which these weapons were made; however, it is conceivable that they could also be made from dense woods or other forms of sturdy material. Even coins could be sharpened and used as shuriken if the need for these weapons arose on short notice.

The early star- and spike-shaped shuriken forms were made of bronze, brass, iron, and other semihardened metals. It was not uncommon to use coins that had been sharpened or pointed as throwing weapons. Subsequent forms were constructed of steel and tempered in much the same fashion as the samurai sword. Today, however, many are constructed from tool steel and alloy of incredible hardness. Commercial varieties are die-stamped, light-gauged metals which are painted with various martial arts motifs. Many experts consider shuriken of this type to be toys rather than "real" weapons.

A "real" shuriken should have sufficient weight, extremely sharp points, and blades that will keep a sharp edge.

An eight-point shuriken (center) has less penetrating ability than the five-pointed type.

Five-Pointed Shuriken

Shuriken Composition

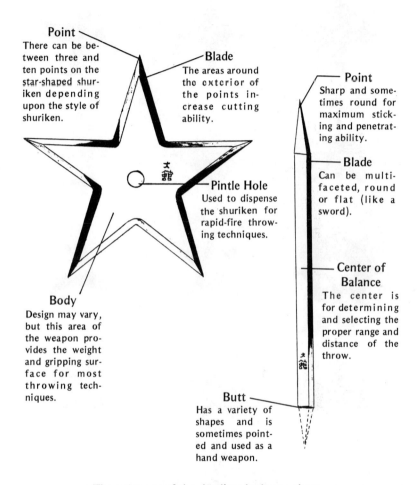

The anatomy of the shuriken is shown above.

These can be purchased from skilled cutlery makers or made if the proper metal-working equipment is available.

The styles presented in this book are among the most popular styles used by the early Ninja, while others are variations on the earlier shapes and designs.

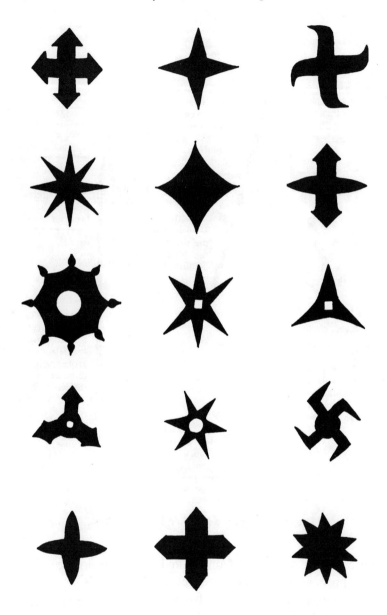

Various shuriken styles are shown here.

5. Gripping the Shuriken

One of the most controversial aspects of shuriken practice involves the gripping of the shuriken.

Since there are so many ways to grip this type of weapon, I will present a variety of methods (and their purposes), which are equally effective—when properly executed. It should be noted, however, that each gripping method has its advantages and disadvantages, many of which will be presented in this chapter. Each shuriken practitioner will inevitably come to his own conclusion as to which method best suits his particular style and stance. It is possible that the serious shuriken practitioner can either discover a unique method of throwing the shuriken that is not covered in this section, or he may utilize a technique combining some of the various grips discovered herein.

The key factors, regardless of gripping method, should be effectiveness and consistency since these qualities are the mainstays of accuracy. A "true feel" for the weapon is also very important so that speed and confidence in throwing can be achieved.

THREE-PRONG GRIP

The three-prong grip is used primarily with the star shuriken. The thumb, index, and middle fingers should be posi-

Three-Prong Grip

tioned in equal increments as closely around the center of the shuriken as possible.

The main purpose of this grip is to create an end-over-end method of delivery. The advantage of this grip style is that the shuriken can be hurled in such a manner so as to enable two or more points to stick into the target, making it very difficult to remove. This "barbed" effect works in much the same fashion as a fish hook. The puncture and penetration produced by the three-prong grip is much more devastating than the conventional one-point penetration technique.

The method of delivery is usually of the overhand or side-arm method.

HORIZONTAL SPIN GRIP

The horizontal spin grip is sometimes referred to as the side-arm delivery because of the horizontal way in which the weapon is hurled.

Gripping the Shuriken 21

The index finger determines the amount of spin applied to the shuriken, while the thumb and middle finger determine the angle of trajectory. Power for this method is generated by winding up the trunk area of the body and coordinating it with the release of the weapon. In the event the shuriken is hurled from the opposite side of the body, the wrist will have considerable effect on the shuriken's spin and velocity.

This grip and method is very effective when the thrower does not have much space in which to throw his weapon.

Horizontal Spin Grip

OVERHAND SPIN GRIP

The overhand spin grip is sometimes referred to as the standard spin grip and is generally used when maximum power and spin are required. This method of release is used quite often when the target is at greater distances from the shuriken thrower.

The index finger of the throwing hand is used to propel the shuriken and determine the amount of spin, while the thumb and middle finger determine the angle of trajectory.

With this grip, the shuriken can be hurled in much the same fashion as one would throw a baseball. With practice, an appreciable amount of curve can be put on the shuriken provided the weapon is not overly weighted. Maximum penetration can be achieved with this method of delivery.

Overhand Spin Grip

UNDERHAND SPIN GRIP

The underhand spin grip is performed in the same manner as the standard spin grip except that the shuriken is released from the underhand position. This method of release is generally used when the target is positioned at an elevation higher than the thrower (such as in a tree).

Velocity is generated by torque which originates in the trunk area of the body, which is then transferred to the arm and wrist as the shuriken is released. Developing proficiency with this method of release will take practice and understanding of the body's mechanics.

Underhand Spin Grip

SLEEVE GRIP

The sleeve grip is a "quick" grip that can be used in the event that one or more shuriken are concealed in a holding device that is mounted up a sleeve. This device results in quick-throwing techniques since the shuriken does not have to be repositioned once it has dropped into the hand from the sleeve.

Most of the power and angle of trajectory are derived from the amount of "flip" generated by the wrist. The effectiveness of this method is displayed when a target is at relatively close range and very little positioning time is available to the weapon thrower.

This method of gripping can be used with the spike styles of shuriken, though it is more effective with the star types.

Sleeve Grip

FOUR-PRONG GRIP

The four-prong grip is used in much the same manner as the three-prong grip except that the ring finger is positioned between the points of the star-shaped shuriken. When this grip is used, the thumb and fingers have to be redistributed equally around the edges of the shuriken so that proper end-over-end trajectory can be achieved. This method is a bit more powerful in its trajectory because a bit more wrist action can be applied to the spin, thereby causing more tearing and ripping effects once it penetrates the target.

Four-Prong Grip

TWO-FINGER TWEEZER GRIP

The two-finger tweezer grip is a relatively simple gripping method. The star or spike shuriken is positioned between the index and middle fingers as far as possible. This is to ensure the best possible grip for controlling the trajectory and amount of spin (or flip) put on the shuriken.

This is a "quick" grip that is used when there is no time to reposition the shuriken or when the other hand is involved in another function. This method is used mostly when shuriken are retrieved from a shuriken pouch that is mounted on a belt or if the shuriken are concealed in a boot or leg pouch.

The force of the index and middle fingers determines the amount of control put on the shuriken, and the flip of the wrist determines the angle and force of the delivery. The direction of travel is determined according to when the fingers are released from the sides of the shuriken.

Two-Finger Tweezer Grip

CLAMP GRIP

The clamp grip is perhaps the most common method of gripping either type of shuriken.

This grip is performed by positioning the index finger and thumb around the shuriken, with a portion of the weapon fitting snugly into the hand. Underhand, overhand, horizontal, and flip methods of hurling the shuriken can be achieved from this grip.

The clamp grip works well for both short- and long-range targets. The direction of trajectory is determined by when the hand is released from the weapon. A tremendous amount of body force—and accuracy—can be put into this throwing technique.

Clamp Grip

THREE-FINGER FAN GRIP

The three-finger fan grip is a very simple method of holding the spike or star shuriken. The thumb is placed on the top portion of the shuriken, while the index and middle fingers are positioned on the lower portion of the weapon.

With a bit of wrist flip, it is possible to generate curve to the star-shaped shurikens. This form of shuriken throwing is especially useful when used against moving targets and when the surface texture of the target area is not vertical to the trajectory angle of the oncoming shuriken. Most throws using this grip will be of the overhand or side-arm delivery; however, this throw can also be employed with underhand delivery.

Power is determined by the amount of wind-up put into the motion, and the trajectory is dependent on when the fingers are released from the shuriken.

Three-Finger Fan Grip

TWO-FINGER FAN GRIP

The two-finger fan grip is identical to the three-finger method except that the middle finger is not employed in gripping the shuriken; the methods of throwing, releasing, and determining trajectory, are, of course, the same. This grip is used when a bit more speed is needed in rapid-fire maneuvers, which is made possible since less finger manipulation is required. This method does not have the range of the three-finger method because less gripping surface on the weapon is possible. What it lacks, however, in distance, it makes up for with speed.

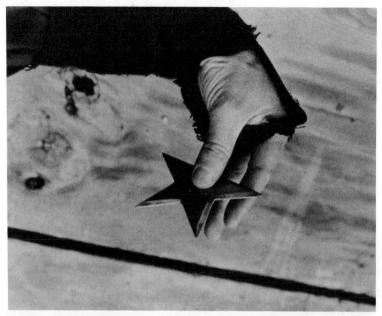

Two-Finger Fan Grip

TRIGGER GRIP

The trigger grip is used primarily with star shuriken and is most effectively used with shuriken having fewer points. Three, four, or five points work most effectively with this type of grip.

The thumb, index, and middle fingers are wrapped around the blades of the shuriken so that they fit snugly between the fingers and the thumb. This type of grip is quite effective when long-range target practice is performed. The power is derived from body involvement, and the trajectory is determined by when the hand is released from the weapon.

This simple gripping method will require considerable practice before accuracy can be achieved. At close range, this method is very effective for deep penetration, though it is very difficult to develop a high degree of accuracy.

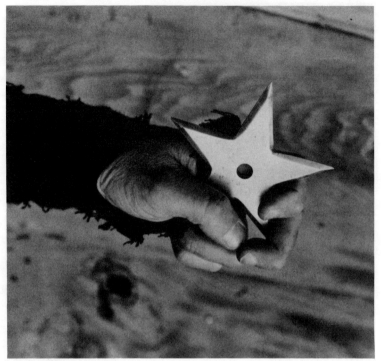

Trigger Grip

PALM GRIP

The palm grip is used almost exclusively with spike shuriken forms.

Either end of the spike shuriken can be positioned in the palm of the hand, depending upon the method of throwing that will be employed.

The spike shuriken is held in the hand so that it is straight and parallel to the extended fingers. The thumb is tucked in in order to hold the shuriken in position. The thumb controls the distance and trajectory of the weapon. Depending upon the experience and expertise of the shuriken thrower, accuracy can be achieved at both short and long ranges.

Palm Grip

THREE-FINGER WRIST FLIP FAN GRIP

The three-finger wrist flip fan grip is another effective way of gripping the shuriken for close-range targets. This method is used primarily with the star-shaped shuriken because of the manner in which the hands are positioned. The thumb is to be placed on the top of the shuriken, while the index and middle fingers are positioned beneath it.

Distance is determined by the amount of flip produced by the wrist and the amount of firmness placed by the fingers on the weapon. This method of gripping is often employed when rapid-fire shuriken throwing is necessary.

Three-Finger Wrist Flip Fan Grip

TWO-FINGER WRIST FLIP FAN GRIP

The two-finger wrist flip fan grip is very similar to the above method except that the thumb and index finger are the only fingers used to grip the shuriken. This method is a bit faster than the three-finger method because of the amount of manipulation required when using rapid-fire throwing techniques.

Since less gripping strength can be employed, this method is less effective at greater ranges than the three-finger method. Power in this method is determined by the amount of wrist action employed, and the flip and trajectory are determined by when the weapon is released from the gripping hand.

Gripping the Shuriken

Two-Finger Wrist Flip Fan Grip

6. Throwing Positions

Stances and throwing positions used with throwing devices, primarily shuriken, will be covered here. Though individual preference will determine the throwing position in a particular situation, the experienced shuriken practitioner will want to be familiar with all the standard and "tricky" throwing positions.

Variations in individual postures and forms will always exist, but the practitioner should be especially aware of the end results achieved with each throwing position described in this chapter. Evasive tactics, surprise, speed, power, rapid-fire throwing, penetration, multi-directional tactics, limited space, and other aspects of shuriken throwing will be emphasized.

It is important to understand the purposes behind each of the techniques displayed in this chapter. A thorough feeling for the body mechanics involved with each throw will enable one to appreciate the unique body positioning required by each stance.

Some of these positions are used for target practice, while others are used specifically for combat purposes. With practice and an innate feel for the shuriken, one will be able to acquire the proper grip, throwing position, and delivery technique to suit the situation.

Natural Stance   Natural Stance Throwing Position

NATURAL STANCE

Most shuriken-throwing techniques can be executed from the natural-stance throwing position. It is used primarily as a posture one assumes before determining whether to take offensive or defensive action. It is also used when maximum awareness and alertness are required.

The feet should be parallel to each other, and they should be about one shoulder width apart. The knees should remain straight, and the knee joint should be free to float so that it will be possible to maneuver in any direction. The knees should *never* become locked when positioned in the natural stance.

CROUCH STANCE

Both legs are bent at the knees and the upper body is in a slightly projected position when one assumes the crouch

Crouch Stance   Crouch Stance Throwing Position

stance. The feet can be positioned in one of the many conventional martial art stances that best suits the shuriken thrower.

This posture generally denotes the amount of combat readiness, whether it be of the offensive or defensive method. It is an excellent position from which to utilize evasive tactics in the event an opponent attempts to initiate the first assault, requiring the shuriken artist to duck, sidestep, or outmaneuver an oncoming weapon. Another advantage of this stance is that many throwing methods can be performed from this position.

HORSE STANCE

The horse stance is best utilized as a defensive measure in order to create the smallest possible target area should an oncoming attack be imminent. Many side-throwing techniques can be accurately delivered from this stance.

Horse Stance   Horse Stance Throwing Position

The feet should be parallel to each other and the knees bent so that quick movement can be initiated. Generally, the width of the horse stance should be twice the width of the shuriken practitioner's shoulders, but this distance varies according to the thrower's taste. Depending on the terrain, one foot may be positioned higher or lower than the other.

FORWARD STANCE

Another favorite fighting stance used by many shuriken artists is the forward-stance throwing position.

The width of the stance, which may vary according to the likes of the artist, is generally about shoulder width, with the lead foot extended forward about one and a half feet. The center of balance should be straight downward from the vertical axis of the thrower's body.

This stance makes it possible to generate a tremendous

Forward Stance   Forward Stance Throwing Position

amount of power in the throw. If the rear hand is utilized to throw the shuriken, the entire hip and upper torso can be put into the throw. When the lead hand is used for flip throws, the body can likewise be put behind the throw.

CAT STANCE

In addition to throwing techniques, close-range kicking maneuvers can be used in conjunction with the cat stance. In martial arts circles, this posture is known for its quick and unpredictable body maneuvers.

Between 90 and 95 percent of the body's weight should be placed on the rear leg, while the remaining weight is lightly placed on the lead leg. The position of the front foot should be pointed in the direction the shuriken artist wishes to throw the weapon.

Cat Stance   Cat Stance Throwing Position

This stance is most often used when multiple throwing techniques are employed and it is necessary for the practitioner to change positions often in order to avoid an oncoming weapon.

BACK STANCE

The back-stance throwing position is used primarily when it is necessary to combine power and speed mobility immediately before or after a throwing technique. Because it permits the artist to be quite powerful at one moment (while executing offensive techniques) and defensive (becoming a smaller target) the next moment, this stance is quite popular with weapon practitioners.

The rear foot should support about 70 percent of the body's weight, while the lead leg will support the rest. The

Back Stance   Back Stance Throwing Position

lead foot should be pointed in the direction of the target, and the rear foot is directed ninety degrees from the front foot.

Many of the flip methods of shuriken throwing are utilized from this position, which is quite effective with short- and medium-range targets.

T-STANCE

Used primarily for target practice, the T-stance throwing position is very similar to the back stance because of the degrees of angle of the feet positioning. It is, however, shorter in length (about one shoulder width), and the weight is equally distributed between the rear and lead feet. The lead foot is pointed in the direction of the target, while the rear foot is ninety degrees from the frontal direction.

T-Stance   T-Stance Throwing Position

Primarily a stationary position, the T-stance is rarely used in actual combat situations because it is difficult to move quickly once the weapon is thrown. Many target-practice competitors use this stance because in-line shoulder trajectory can be employed while throwing the shuriken.

I-STANCE

The I-stance is normally used when consistent accuracy is required in competitive target practice.

A very narrow stance, the I-stance is less than half the width of the natural stance. The feet are placed about four inches apart and parallel to each other. The shoulders are in line with the target, while the feet are ninety degrees from the frontal position. The knees should remain locked so that

I-Stance   I-Stance Throwing Position

the shuriken's trajectory path is not affected by joint turning or twisting.

Since this is an in-line shoulder trajectory stance, only the elbow and wrist determine the power and trajectory of the shuriken. It is consequently used primarily for short- and medium-range target practice and is not a truly effective combat stance.

NOOKED STANCE

The nooked stance is an excellent position in combat situations where multiple opponents are positioned within a 360-degree circumference.

Assume the natural stance. Pivot on the balls of the feet 120 degrees in the same direction while lowering the

Nooked Stance   Nooked Stance Throwing Position

body position until the rear knee touches the rear of the lead leg. Raise the heel of the rear leg. By utilizing the pivoting motion, either to the left or right, it is possible to rotate a complete 360 degrees, thereby making it possible to change directions quickly in the event you are confronted with multiple opponents.

This fighting position works extremely well with rapid-fire shuriken-throwing techniques.

REAR POSITION

Shuriken that are thrown to the rear of your position can be executed from one of many stances. The primary factors involved in proper execution of this delivery method are the amount of pivot necessary to get the body in a position that will permit early observation of the target and proper windup before the shuriken is thrown.

Rear Throwing Position

Knee, hip, and spine flexibility are important factors in determining the effectiveness of this technique. When attempting rotating-throwing techniques, remember that if the knees remain bent, the radius of the turn will be greater than if the knees are upright and locked. Flip or power throws can be accurately delivered from any stances which permit radius turning and body swivel.

Behind-the-Back Throwing Positions

BEHIND-THE-BACK POSITION

Virtually every stance covered in this chapter can be applied to the behind-the-back throwing position, which is employed when the element of surprise is necessary in a combat situation. This method may also be employed when freedom of movement is hindered and space is limited.

Power is generated primarily through hip, shoulder, and wrist coordination, which must be synchronized so that velocity can be increased when the technique is put into motion. This method of throwing and positioning is usually reserved for short- and medium-range targets.

BEHIND-THE-HEAD POSITION

The behind-the-head throwing position requires considerable practice, understanding of body mechanics, and accumu-

Behind-the-Head Throwing Position

lated momentum windup before sufficient speed and power can be generated. When space is limited and freedom of movement is hindered, this is a good position to utilize.

The majority of the force generated for the shuriken's velocity is derived from the elbow and flip of the wrist. The delivery is executed with overhand motion. The most popular grip for star-shaped shuriken used in this throwing method is the standard or overhand spin grip. Failure to achieve a smooth delivery can cause the shuriken to inflict injury to the back of the hand.

Under-the-Leg Throwing Positions

UNDER-THE-LEG POSITION

If you were positioned in a tree or other precarious position where windup maneuvers would give your position away to an adversary, the under-the-leg method would be the method to use. Though it can be executed from virtually any posture or stance, this throwing position is primarily used when freedom of movement and space are limited.

Depending on the angle and position, the power and flip of the shuriken will be derived from the shoulder, elbow, and wrist. Underhand throwing methods are the most effective ways of throwing the shuriken from this position.

BETWEEN-THE-LEGS (REAR) POSITION

When the element of surprise is necessary in order for your efforts to be effective, deliver the shuriken from the

Sitting Throwing Position

between-the-legs (rear) position. It is also useful to do so when an opponent is positioned behind you at short or middle range and when it is difficult to utilize a rear throwing position.

The overhand throwing method would be the primary method of delivery used in this throwing method. When using the star shuriken, use the spin grip method for propulsion. Since the trajectory would be from lower to higher elevations and the vision would be upside-down, considerable practice is required before this method is ever applied in an actual combat situation.

SITTING POSITION

Assume the sitting position so that shoulder, elbow, or wrist action will not be limited. The more limited these areas are, the less power for penetration and less likelihood of

The proper way to grip a five-point shuriken is shown above.

delivering an accurate throw.

One hand is used for supporting the body's weight, while the other is free to flip the weapon in the intended direction. Foot positioning can be an important factor if body weight is to be put behind the throw.

PRONE POSITION

One of the most difficult shuriken throwing positions, the prone position is used primarily when the thrower is concealed in areas where he is required to lie flat on his stomach.

A minimum amount of shoulder motion can be expected, although elbow and wrist motion can be used quite effectively if the shuriken artist is experienced in these position methods. It is possible to achieve additional momentum behind the throw if the feet can be dug into the surface on which one is standing.

In most cases, the overhand delivery method is used when maximum penetration is to be achieved at relatively close range. This method is very difficult to master at medium or long range regardless of the grip or throwing method employed.

7. Drawing the Shuriken

Drawing the shuriken is complicated by the many inconsistent factors involved with dispensing the weapon. Elements to consider include where the shuriken is to be positioned in relationship to the throwing hand, accessibility to the weapon, type of gripping method to be employed, and whether or not the shuriken is to be concealed. The body position or stance the thrower assumes is also an important factor to consider, as are several other technical aspects of logistics regarding the thrower and the ultimate placement of the weapon.

Every possible drawing method that could ultimately be employed with the shuriken will not be presented here— only a clear picture as to the importance and purpose of drawing techniques will be examined. As with most forms of weaponry, it is not necessarily he who is the most proficient with the weapon who wins the confrontation, but who can draw first. This should always be the first consideration when evaluating one's proficiency skills with the shuriken.

It is, of course, more effective to have the throwing hand positioned on the weapon before it is drawn than to reach and fumble with the gripping manipulation before attempting to throw the shuriken.

The Ninja must know how to quietly and gracefully climb stairs.

Whether it be for quick-draw or surprise purposes, a drawing technique should be a dependable and consistent one, since it is better to be a bit slower and reliable than quick and fumbling. If one is extremely quick and fumbles in the drawing attempt, the opponent who may be a bit slower and reliable may ultimately be the most effective of the two. This is most probably true because the slower thrower will usually strike the target first, and that first strike is the ultimate deciding factor when drawing and throwing the shuriken.

The shuriken artist should experiment with various ways of drawing the weapon from different locations, pockets, or weapon pouches until reasonably confident and knowledgeable as to which styles suit specific needs. One should never attempt to use an untried method when drawing the shuriken, especially in critical situations.

8. Trajectory Methods

Trajectory is defined as "the path described by a body moving under the action of given forces." This chapter will deal with the star-shaped shuriken and the given forces and characteristics that affect its trajectory.

The weapon's trajectory is determined in large part by the manner in which it is thrown, though the shuriken's design will also have an effect on trajectory. If the shuriken artist is aware of design variables and their effects, he will better understand the different capabilities of the various shuriken types. The following chart will help to acquaint the shuriken artist with the many variables that influence the shuriken's flight and accuracy. By experimenting with these variables, he will discover many unique ways in which the weapon can be twisted, twirled, flipped, curved, deflected from another object, and rolled as well as other unusual methods the shuriken can be employed.

Stance positions and gripping and trajectory methods will constantly be interacting with each other, and the skilled shuriken practitioner should also realize that most trajectory methods can be used from a variety of stances and throwing positions.

Overhand Trajectory: Step One

Step Two

Step Three

Step Four

Horizontal Side Arm Trajectory: Step One

Step Two

Trajectory Methods 61

Step Three

Step Four

Underhand Trajectory: Step One

Step Two

Trajectory Methods

Step Three

Step Four

Back Spin Horizontal Wrist Flip: Step One

Step Two

Trajectory Methods

Step Three

Step Four

Back Spin Vertical Wrist Flip: Step One

Step Two

Trajectory Methods 67

Step Three

Step Four

Diagonally Downward Trajectory Method: Step One

Step Two

Step Three

Step Four

Diagonally Upward Trajectory Method: Step One

Step Two

Trajectory Methods

Step Three

Step Four

SHURIKEN CHARACTERISTICS

Design
Affects penetration ability, balance, grip ability, trajectory path.

Weight
Affects range, penetrating ability, amount of curve, power loss.

Flight Method
Affects wind resistance, curving ability, penetration ability, deflecting ability.

Curve
Determined by design, wind resistance, amount of velocity.

Power Loss
Determined by velocity, distance of travel, shuriken weight and design.

THROWER CHARACTERISTICS

Range
Determined by gripping method, amount of velocity, shuriken weight and design.

Gripping Method
Affects trajectory velocity, amount of curve and range of weapon travel.

Trajectory Method
Affects velocity, range, curve. Affected by stance position, gripping method, weapon design.

Curve
Determined by gripping method, amount of velocity, shuriken weight and design, wind resistance.

Velocity
Determined by gripping method, amount of wind-up, range of target, amount of wind resistance, stance positioning.

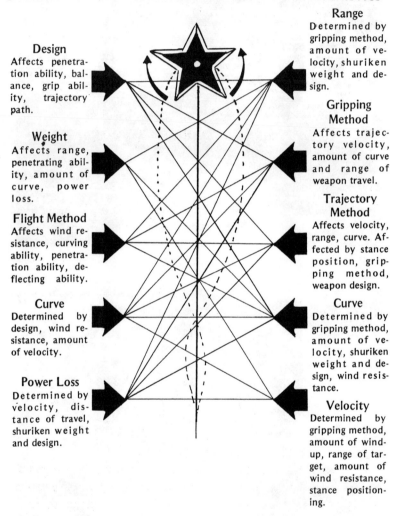

The above data indicates the relationship between the shuriken and the thrower's characteristics which affect virtually every aspect of shuriken trajectory.

9. Penetration Characteristics

It is necessary to understand the penetrating characteristics of the various shuriken styles so that the appropriate one is selected to do the best job for a given situation.

The texture of the target area, the weight of the shuriken, the length and sharpness of its blades and points, and the distance within which the weapon is most effective are all factors affecting penetration. Additional factors to consider are the grip and trajectory of the shuriken.

WEIGHT

The weight of the shuriken should be such that it can maintain a "true" trajectory (not dropping while in flight) for at least thirty feet. The lighter the shuriken, the more likely it can attain a curve effect.

Thickness also plays a role in determining the shuriken's weight, so you should select a thickness that complements the design (for aerodynamic purposes) and provides for a firm and sturdy grip.

Many of the die-stamped "toy" shuriken are made of very thin gauged metal, and as a result the shuriken does not possess sufficient weight to be propelled greater distances or to penetrate sufficiently to stick consistently to the target.

Most shuriken, however, have a thickness of at least three-sixteenths of an inch, which provides the necessary weight to project the shuriken a minimum distance of thirty feet.

NUMBER OF POINTS

The style, diameter, or length of the shuriken will ultimately determine the number of points the weapon will have. Generally, the more points, the less likely the shuriken will achieve adequate penetration to be used proficiently as a weapon. The more points there are, the less space exists around the shuriken for extended blade surface. As a result, the sticking surface on a target is limited upon impact. The amount of penetration surface will vary, depending on the size of the shuriken.

If a shuriken is six inches in diameter as opposed to two inches, the length of the six-inch blade would be longer even though both sizes have the same number of degrees intersecting at the connecting points between the blades. The longer points would give the larger weapon greater penetration capability.

The sharpness of the points is critical as well. Theoretically, you can have a deeper-penetrating shuriken that is small, provided the weight and sharpness are of better quality than those of a larger one.

TEXTURE AND GRAIN

The texture and grain of the target area must be determined and thoroughly evaluated if the maximum penetrating effect is to be obtained.

"Texture" refers to the hardness or density of an inanimate target; "grain" refers to the direction in which the growth rings run in wooden targets. In many cases where a shuriken is tempered or case-hardened, texture and grain are insignificant, provided that sufficient velocity is applied to the trajectory of the shuriken.

The shuriken artist should always strive to strike a wooden target so that the weapon penetrates parallel to the grain of the target surface. Maximum penetration is then ensured, provided that ample force is used for the texture of the target surface. If the target surface is tilted, the shuriken will strike with a glancing blow, and it may be necessary to use additional force or another style of shuriken having fewer points. Exerting excessive force when throwing a shuriken can only hamper the ability to maintain a consistent throwing style.

Regardless of the texture or density of the target area, it is important that a very sharp point and finely sharpened blade area are maintained.

VELOCITY

There are four ways to generate power or velocity in the trajection of the shuriken, and with proper technique, they can gain maximum effectiveness for the amount of energy expended.

First, brute force can be used, winding up the body and suddenly releasing the weapon. In most cases, however, a consistently accurate delivery will not be achieved by this method.

Second, arm force and wrist action can be employed to put tremendous momentum on the weapon. In this way, a consistently accurate delivery can be achieved.

Third, hip and proper muscular sequential trajectory can be employed, thus combining brute force power and wrist and arm force in one clean, accelerated maneuver. This method should be practiced for maximum penetration and accuracy. Because the hips are slower than the upper body and arms, it is necessary to develop a smooth, circular, forward hip motion. Coordinate the hips with the upper body and arm so that they complement each other in order to increase the accelerated velocity of the shuriken.

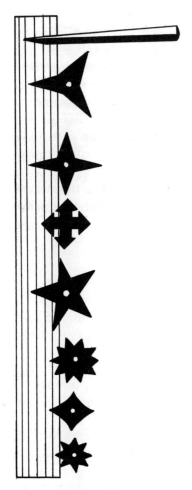

1-POINT STYLES—Penetrate most target surfaces extremely well.

3-POINT STYLES—Penetrate well provided the weapon is of sufficient weight for its size.

4-POINT STYLES—Penetrate well provided the weapon is of sufficient weight for its size.

4-POINT STYLES—Penetrate moderately well if the weapon is of sufficient weight for its size and blades are very sharp.

5-POINT STYLES—Penetrate extremely well provided the points and blades are sharp and the weapon is of sufficient weight for its size.

10-POINT STYLES—Do not penetrate well because the points are positioned too close even when they are of adequate weight.

4-POINT STYLES—Do not penetrate well, but can cause severe damage by cutting.

8-POINT STYLES—Penetrate the least because the points are positioned very close together and usually lack sufficient weight.

The penetration characteristics of shuriken vary according to the number of points in a particular style.

Your body will develop a feel for determining when adequate power (velocity) is or is not being generated. The smoothness of the windup will feel like swinging a baseball bat and then making a connection with a ball that scores a home run.

You may not be able to generate this velocity using certain gripping methods due to body posture or a problem with flipping the shuriken. Do not be alarmed. Just use the stances and postures that best suit your need for maximum penetration. The release of the weapon should be early enough to ensure that body momentum has not been diminished before the shuriken is airborne.

METHOD OF TRAJECTORY

As previously mentioned, trajectory method is crucial to achieving maximum penetration with the shuriken. The method of trajectory includes stance, body positioning, and gripping technique. You will be able to determine the proper method of trajectory by feel, comfort, balance, timing, release technique, sequential muscular augmentation, and distance from the target.

DISTANCE FROM TARGET

The final determining factor for ensuring maximum penetration of the shuriken is the distance from the target. Because velocity decreases with distance, it is necessary to generate enough power (velocity) on the weapon before it is released from the hand. After release, it will steadily lose velocity. You will want to ensure that a true trajectory is maintained between you and the target and that no drop occurs with the in-flight shuriken in order to improve the accuracy of your throw and to increase the weapon's penetration.

If too great a range exists between you and the target so that maximum controlled power cannot be achieved, it will become necessary to rely on luck and haphazard methods and techniques. Avoid these at all costs since they will have a negative effect on both your accuracy and the confidence you have gained in your shuriken skills.

10. Sighting Methods

Sighting methods are used in conjunction with trajectory and release elements of shuriken throwing. Depending on his individual preference, the shuriken artist can use any one of a number of sighting methods when using the shuriken. All of these methods have one thing in common: a point where the line of sight between thrower and target and the travel path of the thrown shuriken intersect. Regardless of the point of release, there will always be a point along the travel path of the shuriken where it will intersect and converge with the line of sight. The exact position of this point will depend on the distance between thrower and target and the thrower's individual technique.

Usually, increased consistency, accuracy, and control of flight direction can be obtained if the intersecting point is closer to the thrower than to the target. This is because there are more reference points for the shuriken thrower to relate to at close range than at a distance. Many shuriken throwers find that two feet past the reach of the extended throwing hand is an excellent point for the shuriken travel path and the line of sight to intersect. Again, it should be mentioned that this distance may vary from thrower to thrower, depending upon individual style.

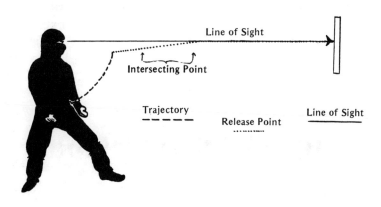

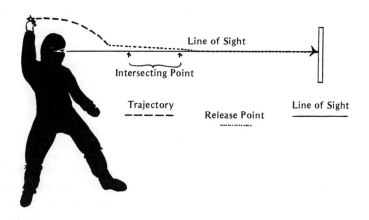

By experimenting, it will be possible for each thrower to determine the proper distance between himself and the target, thereby ensuring consistency and accuracy. Two trajectory methods are shown above: underhand (top) and overhand (bottom).

All throwing and sighting methods demonstrated in this chapter can be performed from any of the stances and body positions described in this book. The serious shuriken artist should become adept at utilizing each and every stance with all the various trajectory methods until total understanding of the sighting methods is achieved. It is essential that the thrower develop reference points between the extended arm, the release point of the weapon, and the fixed body posture if consistency and accuracy are to be developed and maintained. Haphazard throwing with little consideration for fixed reference points will lead only to lucky throwing and inconsistent trajectory of the shuriken. In shuriken fixed target competition, there will be very little room for luck.

LINE OF SIGHT

The line of sight plays an important role in establishing a consistent and accurate throwing method. Regardless of stance positioning or trajectory method, the throwing arm must have a clear and unobstructed path of travel and be able to intersect the line of sight between the thrower and the intended target.

For maximum consistent accuracy, the thrower must have fixed positions of reference in relation to the line of sight in order to coordinate an organized and methodical release.

INTERSECTING POINT

The intersecting point is the point at which the shuriken converges and travels in line with the thrower's line of sight to the target. As mentioned above, this point is generally about two feet past the thrower's extended arm. A point of intersection at a distance greater than two feet may decrease the thrower's control because he will have fewer fixed reference points from legs to trunk, trunk to body, body to shoulder, shoulder to arm, arm to hand, and hand to weapon to incorporate into the shuriken's trajectory. It tends to make the thrower feel he has no control of the direction of travel of the shuriken, and makes him less likely to be consistent in his throwing method.

RELEASE POINT

The time when a thrower releases the shuriken may vary considerably depending upon his expected results. If the release point is delayed too long beyond the intersecting point, there will be a loss of velocity in the throw. It will then be difficult to control the direction of the shuriken. If the release point is too early into the trajectory of the weapon, however, insufficient power will be generated for longer range targets. By experimenting, it will be possible to discover the proper timing so that the shuriken will intersect the line of sight at about two feet from the extended throwing hand. The release point will vary with the shuriken's configuration and the type of propulsion and spin put on the weapon. The fixed body reference points will be of great help in assisting the thrower to determine the proper time to release the shuriken.

11. Accuracy

Accuracy with the shuriken is difficult to achieve because of the various skills required of the practitioner. One method will seem orthodox to one individual and unorthodox to another. Personal preference for the type of shuriken, feel, balance, weight, distance, gripping technique, trajectory method, stance positioning, ability to concentrate, innate feel for release of the weapon, size and distance of target, judgment of weapon velocity, wind-resistance factors, target texture, proper power/distance ratio, eye-hand coordination, sufficient weight/range ratio, individual style preferences, ability to perform under pressure, and other unforeseen qualities unique to each shuriken artist play an important role in determining the accuracy of the thrower.

It is not the purpose of this chapter to dictate the proper method for achieving accuracy for each shuriken artist, but rather to present the subject so that the thrower can make the choices proper to his particular style.

Each shuriken artist should strive for consistency regardless of his throwing approach. If consistency can be achieved, accuracy will inevitably be the reward of the thrower's dedicated efforts. Each individual will take a different route and develop a unique way of performing each of these skills. In

POSITIVE EMOTIONAL FACTORS	POSITIVE PHYSICAL FACTORS
CALMNESS	PROPER POWER
CONFIDENCE	FAMILIARITY WITH WEAPON
CONCENTRATION	GOOD EYE/HAND COORDINATION
CONSISTENCY	CONSISTENCY IN TECHNIQUE
OBJECTIVE	PROPER METHODOLOGY
PATIENCE	GOOD POSTURE
DELIBERATENESS	PROPER GRIPPING METHOD
RESERVE	PROPER BODY POSITIONING
PERSEVERANCE	PROPER POWER TO DISTANCE RATIO
TEMPERANCE	PRACTICE
PERSISTENCE	PROPER RELEASE TIME JUDGMENT
MODERATION	PROPER WEAPON
EXPERIENCE	KNOWLEDGE OF WEAPON CAPABILITIES

If any of the above qualities are omitted from the practitioner's regimen, it will be very difficult to develop consistency and accuracy with the shuriken.

Truly prepared, the Ninja has a shuriken in each hand.

all probability, no two shuriken artists will perceive any of them in exactly the same way.

ACHIEVING ACCURACY

Achieving and maintaining accuracy when using the shuriken requires a tremendous amount of time and patience. One must set a goal and not be diverted from that goal until it has been reached. One's standards must be high, but one must never be discouraged by shortcomings, realizing that they are only temporary and that progress will be made if effort toward the goal is continuous.

HUMAN FACTORS

The "human factors" are a combination of both physical and emotional human qualities. When negative in nature, they can have a devastating effect on one's ability to develop accurate shuriken-throwing techniques. If the shuriken artist is aware of their existence, he will be better able to deal with their effects on his throwing accuracy.

Many negative emotional factors can be eliminated by applying a personal method of positive thinking, while others can be eliminated by developing a meditation method that suits the individual.

Negative physical factors, on the other hand, can be eliminated for the most part by learning the proper gripping, stance, and trajectory methods presented in this book, and understanding the reasoning behind them. There is no substitute for practice, and this is one of the most important physical factors determining the ability to achieve accuracy.

NEGATIVE EMOTIONAL FACTORS	NEGATIVE PHYSICAL FACTORS
Tension	Excessive power
Fear	Lack of familiarity with weapon
Anxiety	
Nervousness	Poor eye/hand coordination
Lack of confidence	Inconsistency in technique
Lack of concentration	Poor methodology
Shyness	Poor posture
Inferiority complex	Improper gripping method
Moodiness	Improper body positioning
Excitement	Lack of power/distance
Inconsistency	Lack of practice
Temper	Improper release time judgment
	Improper weapon for job

EMOTIONAL FACTORS

Consider the negative effects of the following emotional factors.

Tension. Tension will affect every aspect of the thrower's technique, both mentally and physically, regardless of his prior experience and expertise. Physical exercise before competition helps alleviate tension.

Fear. Fear often causes the shuriken artist to become inconsistent in his throwing ability. Most shuriken throwers perform very poorly until they are comfortable with the environment and situation.

Anxiety. This negative emotional quality usually begins acting upon the thrower prior to serious competitive encounters. Relaxation and meditation help to relieve anxiety.

Nervousness. Nervousness affects most of the physical aspects of the thrower's art. Relaxation, meditation, and a positive attitude help eliminate this emotional quality.

Lack of Confidence. The thrower who lacks confidence generally lacks training or practice in throwing. Additional practice generally helps eliminate this emotion.

Lack of Concentration. Poor concentration negatively affects throwing consistency and, as a result, throwing accuracy. One should develop a good method of employing the tunnel-vision concept.

Shyness. Shyness in itself does not usually affect physical performance, but it can contribute to an inferiority complex which is detrimental to the thrower's physical prowess. A shy person should strive to develop strong concentration methods.

Inferiority Complex. An inferiority complex is very common with shuriken artists with little competitive experience. Experience and a good understanding of basic elements of the physical aspects of the throwing art help to correct this problem. A positive attitude will also help develop sound throwing practices.

Moodiness. Changeable emotions generally make the throwing artist inconsistent in both practice sessions and competitive situations, negatively affecting his accuracy. Strong and disciplined training schedules help eliminate this negative emotional quality.

Excitement. Excitement can be either positive or negative, depending on when it affects the shuriken thrower. During a competitive encounter it may cause overconfidence, possibly upsetting consistency. Before a competitive event it can decrease the thrower's ability to concentrate during the competition. One should strive to develop concentration, using a strong tunnel-vision method.

Inconsistency. Inconsistency contributes to emotional and physical problems when throwing the shuriken. Positive thinking, confident attitude, strong concentration and objectivity are necessary to overcome inconsistency. Continuous practice and a regular training schedule are also of great benefit.

Temper. A calm and confident attitude will help eliminate this negative emotional quality. Temper can affect other emotional qualities as well as physical emotional qualities. The combined effects will influence consistency, and ultimately throwing accuracy. Remain calm at all times and do not linger on past competitive performances.

PHYSICAL FACTORS

The following physical factors can adversely affect a shuriken-throwing technique.

Excessive Power. Excessive power in a throwing method works against a shuriken thrower's consistency and accuracy. One should strive to use appropriate power for the distance over which the shuriken is to be thrown. This is accomplished through experience and familiarity with the shuriken and trajectory method.

Lack of Familiarity with Weapon. This negative physical factor results from insufficient training and a resultant lack of capability and understanding of all aspects of shuriken throwing. This problem can be eliminated with practice and experimentation with the trajectory possibilities of the weapon.

Poor Eye/Hand Coordination. If the shuriken thrower is in good physical health and there is no medical reason for poor eye/hand coordination, a lack of good gripping, positioning, or trajectory skills can be its cause. These three elements are interrelated, and it is essential that the thrower understand their proper sequential relationship; otherwise, inconsistency and inaccuracy will result.

Inconsistency in Technique. This negative physical quality is often responsible for inaccurate trajectory. In competitive situations, one should never attempt to experiment with throwing style. This problem can be overcome by thoroughly understanding and practicing methods of throwing that have in the past been successful.

Poor Methodology. Poor methodology is defined as haphazard gripping, stance positioning, and trajectory. Inconsistency, erratic performance, and insufficient skills generally result from this lack of concern for proper method of getting the shuriken from point A to point B.

Poor Posture. Poor posture will affect every physical aspect of shuriken throwing. A good understanding of balance, center of gravity, and upright body positioning is necessary if consistent aim and trajectory are to be achieved. Poor posture can be eliminated by practicing correct posture in front of a mirror.

Improper Gripping Method. Although there are many ways to grip the shuriken, the thrower must be familiar with them all in order to determine which suits his particular need for a specific purpose. If improper gripping methods are utilized, inconsistency in releasing the shuriken from the hand will result, thus marring overall performance.

Improper Body Positioning. There are many stances and positions that can be assumed when preparing to throw the shuriken, and it would be impossible to dictate the proper position to suit the needs of each individual.

A thrower's posture determines the amount of power or velocity of the weapon as well as the sighting to the target. It also controls the reference points for the release of the weapon.

These criteria must be met, regardless of the stance used, if total and consistent accuracy is to be achieved.

Lack of Power/Distance. If a shuriken thrower lacks the

power to correctly hit the target, he will never become accurate with his trajectory. If the power of the trajectory feels comfortable, it may be necessary to select a lighter shuriken of the same shape and design. If this is not the case, it may be necessary to increase the velocity behind the weapon and adjust to a more powerful throwing technique.

If these remedies do not serve the purpose, the stance and positioning from the target may need to be modified so that more body and mass can be implemented into the trajectory method.

Lack of Practice. One must be dedicated to regular practice habits, striving to set goals for expected achievement in every session and then continuing each session until these goals have been reached. If this form of practice is implemented, the practice sessions will be not only productive, but fun as well. Lack of practice will negatively affect accuracy and consistency.

Improper Release Time Judgment. If trajectory and methodology are correct and inconsistency is still occurring, the problem may be the time at which the shuriken is released from the hand. Only by experimentation and close observation can the thrower learn the proper time to release the shuriken.

If no consideration is given to this important aspect of shuriken throwing, one will be inconsistent and inaccurate in his throwing style.

Improper Weapon for Job. If you have studied the weapon characteristics and styles thus far presented, you will have a good understanding of which type of shuriken is best for a particular target. Since every thrower will have his own definitive style for throwing the shuriken, he will place a great deal of importance on the weapon once he has gained a degree of accuracy and consistency. He may be willing to upgrade the quality and style of his weapon to achieve a more critically balanced or weighted weapon. If his ultimate goal is accuracy, every aspect of throwing, including the quality of the shuriken, will help him reach his goal.

12. Concentration

Concentration, the ability to center one's energy and not be distracted by outside interference while performing a specific task, is perhaps one of the least understood ingredients required by the shuriken artist.

Tension, fear, anxiety, lack of confidence, excitement, the unknown, pressure—these are but a few of the distractions that affect the thrower, whether in competition or in combat situations. One must develop a method of eliminating such distractions.

One's desire and enthusiasm must be greater than one's interest in outside activities before one can begin to develop concentration. This desire must be intense enough that mental determination will block out distractions that would affect the performance of the shuriken thrower.

TUNNEL-VISION APPROACH

There are several ways to eliminate distractions and develop a concentration that can be recalled when the situation demands it. One of the most effective ways I have discovered is the tunnel-vision approach in combination with a strong desire to achieve superlative results.

The tunnel-vision approach is used primarily to center your vision and make you aware of your desired objective and the relationship between yourself and your target. The strong mental desire is achieved by reenacting, or seeing yourself perform the throw before you actually throw the shuriken.

Admittedly, to do so requires a vivid imagination at first, until you become familiar with this fantasy approach to drawing full mental power to the desired objective. When applied in conjunction with the tunnel-vision method of centering the visual energy, real concentrated energy can be applied to your shuriken-throwing skills—with almost immediate positive results.

To achieve maximum results from this method of developing concentration, it is necessary that you become thoroughly familiar with the weapon and acquire a consistent throwing method so that the trajectory is physically comfortable, and you and the shuriken seem as one. Accuracy is not necessary at first, provided that the method of delivery is consistent.

Using this concentration method, it will be possible to change consistent inaccuracy into consistent accuracy. Your ability to concentrate, however, must be so strong that noise, talk, or other distractions do not take you away from your thoughts. At first this will be difficult; you may become so involved with the concentration process that you find yourself thinking about the process and lose your intended objective. Do not, however, be discouraged. Keep practicing until you have mastered this method of developing a strong concentration, and mastery of the shuriken will eventually be achieved.

To develop full concentration, all distractions must be eliminated, and mental and physical energy must be centered into a focused target. This method of developing concentration requires practice until you are capable of seeing only the desired target and you can automatically focus your energies and attention to gain the desired results.

13. Spike Shuriken Trajectory

Spike shuriken have only one or two points, and there are essentially two styles of throwing them: the *direct* method and the *flip* method. For close-range targets, the direct method seems to be most effective for the majority of shuriken practitioners, while the flip method works better at greater distances. It is a matter of personal taste as to which method will be suitable for intermediate ranges, but practice is in any case necessary to give the thrower the needed information to determine which method works more effectively for him.

The direction the pointed end of the weapon is facing before the throw is attempted must be considered. If it is a double-ended spike shuriken, this will make very little difference. However, the thrower must be skilled at both trajectory methods if the proper choice is to be made for a particular distance.

All throwing positions and stance forms can be applied to both the direct or flip methods of shuriken throwing, and the serious practitioner will make every attempt to master and discover the many possibilities that can be derived from each.

It should be mentioned that body alignment and timing are two other important aspects of these trajectory methods.

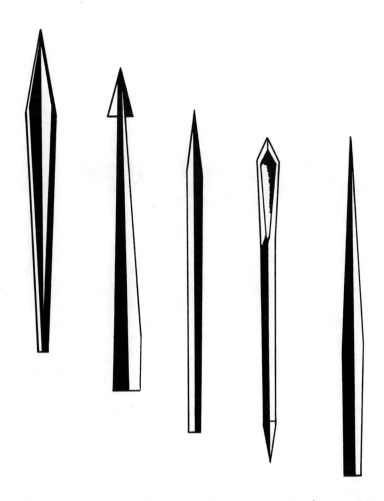

These popular spiked shuriken styles vary in weight (two to fifteen ounces), length (five to eight inches), and width (three-eighths to one-and-a-half inches).

To ensure maximum consistency and accuracy, knowledge of body alignment and timing should be developed and practiced continually until favorable results are obtained.

DIRECT METHOD

When the thrower is in alignment with the target, the shuriken travels along a straight and direct path to its target.

FLIP METHOD

The shuriken flips three hundred sixty degrees before it strikes the target in this method, which is a bit more difficult to master.

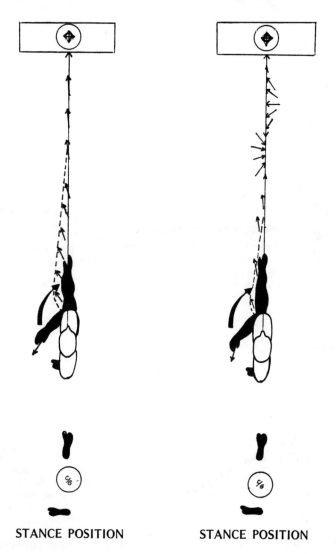

STANCE POSITION   **STANCE POSITION**

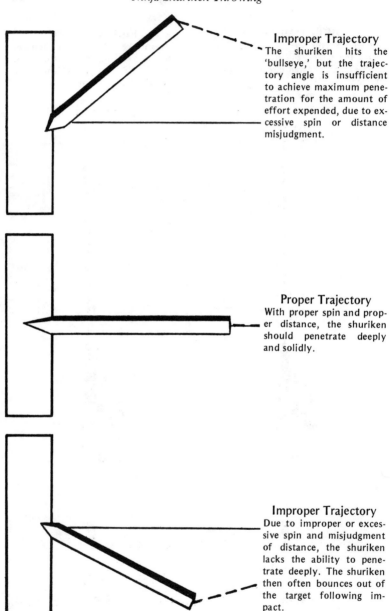

Effects of proper and improper trajectory for the spike shuriken are illustrated above.

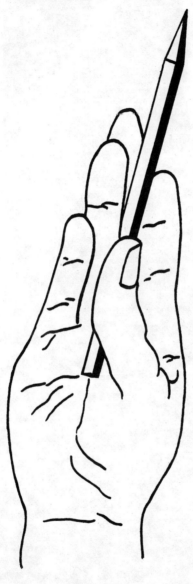

The butt end of the single-point spike shuriken should be placed snugly and securely in the center of the palm. The fingers should wrap inward, while the thumb is placed firmly on the flat side of the shuriken blade.

14. Fighting Tactics

In any form of tactical warfare, it is necessary to know your adversary and his capabilities, and shuriken fighting is certainly no exception.

The methods covered in this chapter should be thoroughly understood so that they can be employed instinctively and spontaneously in the event their implementation becomes necessary. In many cases, the ability to react quickly and expertly will be the factor that determines the outcome in an armed or unarmed combat situation.

RANGE (DISTANCE)

In armed confrontations, it is usually the one with superior capabilities at greater range who has the distinct advantage. This simply means that the one with the weapon that has a greater shooting, firing, or throwing range can outdistance an opponent with a weapon with a shorter range capability. Since the shuriken can be used at close or short range and its range and velocity are determined by the thrower's physical ability, its effectiveness would be limited when used against firearms.

In extremely close-range encounters, many empty-hand methods of martial arts can be applied in combination with the shuriken techniques. If, however, one is at a greater distance and these additional skills cannot be employed, it will become necessary to utilize other tactics. It is always necessary to determine the effective range of your fighting capability and then determine the effective range and type of weapon your opponent has before other methods and tactics are applied. It should be kept in mind, however, that if obstacles, barriers, and structures (such as houses and cars) are between you and an adversary with a superior range weapon, the effectiveness of his weapon is diminished considerably.

AN ARMED OPPONENT

An armed opponent, especially one who has made his intentions known, can be the most deadly of all. If you are only armed with the shuriken, it is highly advisable to put as much distance as possible between you and your adversary in the shortest possible time.

Since your adversary may be armed with a pistol or other type of firearm, it is also advisable to keep as many barriers or obstacles between you and him as possible. By using stealth techniques, it is feasible to weave or maneuver in an erratic or unpredictable path while performing these evasive tactics. If an armed adversary is encountered during the darker hours, it is possible to be a bit more creative and unpredictable in your methods of defense.

If the range is extremely close and you can use empty-hand defensive measures, it is not recommended to make an unsure or false move that may give your intentions away. Remain calm, use diversionary tactics or verbal distraction methods, and, if the risk is not too great, try to confiscate your opponent's weapon. It should be remembered that at extremely close range, many shuriken can be used like knives, but the shuriken must be concealed until it is put into play. In close-range encounters when the element of surprise is on your side, only the vital areas of your opponent's anatomy should be considered. Use the shuriken as you would a knife, but never throw it at the opponent.

Since many of the shuriken serve as nuisance weapons due to their limited penetration capabilities, it is also advisable to have more than one available. Never attempt to use your last weapon against an armed opponent, because in throwing the shuriken you become disarmed and your opponent may end up with your weapon in addition to his own. Throws should only be attempted when you possess numerous shuriken.

In most cases, regardless of the range, the gun-bearing opponent has the advantage of firepower, range, ammunition, velocity, and "kill-power." All methods of utilizing the shuriken should be as secretive and inconspicuous as possible.

AN EMPTY-HAND OPPONENT

Many of the tactics detailed in this section apply more appropriately to the opponent armed with a comparable weapon or to the unarmed opponent who is physically larger or more aggressive than yourself. If one has black belt experience in any of the Oriental martial arts, empty-hand tactics will certainly work well in conjunction with the ones presented in this section.

If one lacks martial arts tactical skills, this book will be of some value in understanding the concepts and methods necessary to achieve the confidence to deal expertly and proficiently with the unarmed opponent.

DISTRACTION

The art of distraction is common in both empty-hand and armed fighting techniques. Distraction methods include verbal, visual, or suggested forms or maneuvers that are employed to give you the advantage over an opponent. The purpose of these methods is to break your opponent's concentration long enough for you to take advantage of his inattentive state. In many cases this distraction time may be very short, and you must be willing to react spontaneously when this lax time exists.

Numerous distraction techniques can be employed in fighting. With a bit of imagination, it is possible to create a method of distraction suitable for virtually every situation that may arise in armed or unarmed combat situations.

This fighting tactic will work very well with other forms presented in this chapter.

ELEMENT OF SURPRISE

The shuriken artist should realize that there are numerous surprise tactics that can be practiced to better one's position against an armed or unarmed adversary. The element of surprise should always be at the forefront of the tactician's mind and should always be one of the first tactics used.

Especially in warfare, the edge gained from this method is all that is necessary to ensure victory, provided the surprise maneuvers are precisely and expertly executed and immediate follow-up is performed. One should think quickly and be cunning when devising methods of surprise and never fail to take advantage of any opportunity which arises.

THE KIAI

The Kiai, the sudden explosive yell heard when an offensive or defensive technique is executed by a karate or Kung Fu practitioner, is very useful to the armed or unarmed combat strategist.

The purposes of this yell are (1) to surprise an opponent, causing him to become confused or disoriented long enough for the tactician to take advantage of the situation; (2) to generate more body power in unarmed confrontations while also using proper breathing and muscle coordination; and (3) to prepare the body by forcing the air out and tensing the muscles, so that if one is struck, the body is prepared to take the shock of the blow, thus sustaining less injury than would otherwise occur.

This fighting tactic is primarily used at very close range in order for the strategist to achieve maximum follow-up results. The Kiai works extremely well with other close-range fighting tactics.

VERBAL INTIMIDATION

Verbal intimidation is considered a form of distraction, but it can be a specific art itself when used in either armed or unarmed confrontations.

The intimidation can be either positive or negative, and the tone in which it is communicated also serves a useful purpose for the strategist.

Body gestures, when incorporated into the verbal intimidation method, will enhance the technique. Intimidation must be convincing if it is to be used to force an opponent to react in a manner that he had not originally anticipated.

The purpose of this fighting tactic is to make your opponent react in an irrational manner by working on his weakness to suggestion or by arousing anger.

Since there are so many verbal intimidations that can be applied to a situation, each individual will certainly find ones to suit his particular fighting style. The important thing to remember is that the method must be convincing, and the desired results must be obtained if they are to be effective in combat situations.

This method works well in conjunction with other fighting tactics presented in this chapter.

OBJECTS

The use of objects has always been a favorite fighting tactic that has earned the reputation of being one of the dirtiest fighting methods, regardless of whether it is in armed or unarmed confrontations. *(The use of objects* refers here to objects other than the human body used in unarmed combat, and objects other than the weapon used in armed combat.)

It will not be the purpose here to list every usable object, but to give examples of techniques in which objects may be used.

In unarmed confrontations, any object can be picked up and used directly or thrown at the adversary. When the object is permanently fixed, the opponent can be maneuvered so that the object is directly behind him. Then, as a forceful empty-hand assault is implemented, the opponent will back into or over the object, losing his balance, becoming confused, and possibly tripping. This will give you the advantage if you quickly follow up. Delay will allow your opponent to regain his balance and composure, and quite possibly make him aware of your motives.

This fighting tactic works extremely well with other methods presented in this chapter.

HIT AND RUN

Regardless of whether one's tactical ploys are offensive or defensive, hit-and-run fighting tactics offer the shuriken artist two distinct advantages over remaining in one position before, during, or after the maneuver's execution.

First, hit-and-run tactics reduce the chances of presenting an opponent with a sitting target. If one remains stationary too long, an opponent will have ample time to establish and initiate his attack or counterattack, thus increasing his chances of using effective ploys.

Second, moving before, during, or after an assault or defensive ploy will create confusion and apprehension in your opponent. This enables you to be totally unpredictable in the eyes of your opponent in the event the confrontation is sustained for any period of time.

The hit-and-run tactic works equally well against armed

and unarmed opponents and should be used with other tactics described herein.

FAKES

There are many types of fakes that can be used quite effectively against the armed or unarmed opponent. Many martial-arts-oriented practitioners confuse the "true" fake with other tactics that are described in this chapter. I define the fake as a technique of feinting executed so skillfully that the opponent actually thinks he sees an action which was never really executed. If the opponent, seeing the tactician's body begin a movement, assumes a particular technique will be executed and reacts accordingly though the movement or technique never really develops, then this is considered the true fake.

This method will work very effectively in close-range encounters and moderately well at short ranges when using the shuriken, provided the shuriken thrower has developed a good sense for faking throwing maneuvers and can get the recipient to flinch or react in a manner that will draw his defenses away from the real target.

This fighting tactic will work extremely well in conjunction with other close-range fighting tactics.

HALF-COMMITMENTS

Half-commitments are often confused with the true fakes covered in this chapter. It should be realized from the outset that the half-commitment involves the actual execution of the intended technique or maneuver against the opponent. If the half-commitment is successful, he will react in the same manner as if a fake were used. The main difference between the fake and half-commitment is that the latter is directed toward the opponent, then retracted before or at the halfway point to its destination.

Because the purpose of the half-commitment is exactly the same as that of the fake, to make the opponent flinch or react in a specific way, this will serve as an alternative method in an extended confrontation. This method works most effectively when one is dealing with an unarmed opponent or an armed opponent at extremely close range.

The half-commitment serves as an alternative fighting tactic and works well in conjunction with many of the close-range tactics explored herein.

FULL-COMMITMENTS

No difference exists between a full-commitment and the actual execution of a method or technique when it is applied in an armed or unarmed confronation. It is emphasized here because it is used on occasion as a method of faking or misleading the opponent. In this case, the actual movement or technique is executed to the full extent of its range and, after impact or near-impact is achieved, the striking weapon or hand or foot is returned to its original position.

As with the fake and the half-commitment, the primary effect to be achieved with this tactic is a flinch response or emotion from your opponent. The retraction of the technique is the ultimate quality of this fighting tactic.

This method applies more to unarmed combat than to armed conflict strategy unless the weapon can be returned to the tactician's hand. Even in such cases, this method is rather rare. The full-commitment will work extremely well as an alternative to fakes and half-commitments in addition to actual execution of a tactical ploy or technique.

This method works well with other empty-hand strategies included in this chapter.

DRAWING

This fighting tactic works well against both unarmed and armed adversaries. Since there are many methods, some unique and less obvious, some obvious and more common, it should be the intention of the serious shuriken artist to pursue the many avenues to discover the specific method to suit a particular need.

Drawing is essentially a method of forcing an opponent to come closer to you. This can be achieved in many ways. As an example, if you are engaged in an unarmed confrontation with a larger opponent whose range for kicking and punching is larger than yours, you can create openings in your defense to force him to initiate an attack. He is thus brought closer to you for the possibility of counterattacking.

It is important to realize that you need to remain alert when using this tactic so that your recovery is well planned and your opponent cannot overrun your position. In armed, close- and mid-range positioning, it is necessary to create enough of an opening to force your opponent to expend his weapon or ammunition without putting yourself in grave danger. If you are outdistanced in an armed conflict, this method can be used to draw your opponent into your weapon range.

This tactic will work with the other fighting tactics I have presented in this chapter.

DECEPTIVE CLOSING

The deceptive closing fighting tactic is another strategy that will work very well against armed and unarmed adversaries when the distance of several inches makes the difference between success or failure.

The success of this tactic is based on the opponent's reaction to an optical illusion; the leading hand or closest part of your anatomy remains the same distance from him, giving a false impression of distance. In the meantime, the remainder of your body moves in closer to his position.

To explain briefly, when one is within a close and critical range to the opponent, the opponent will remain relatively calm provided his safe range has not been entered. By discovering your opponent's safe range by crowding his position, you will certainly get a response from him. If you remain out of that range, the opponent tends to think he controls the distance and can select techniques of attack or defense of his own choosing.

As you begin an overt movement with the lead part of your body (either up and down, in and out, or side to side), the feet gradually move forward the distance necessary to make good your assault. The hands or lead portion of your body are simultaneously retracted the same distance that the feet have moved forward. Optically, your opponent will perceive he has maintained his safe range while in actuality you have closed the necessary distance to make your assault effective.

BREAK RHYTHM

The break-rhythm fighting tactic can be used at close range in unarmed confrontations and, in some cases, in close-range situations where the opponent wields a knife or other handheld weapon (excluding a firearm).

Most fighters have a style of movement that suits their body build or a preference for a particular type of empty-hand martial art with which they have had prior experience. The break-rhythm fighting tactic is designed to upset an opponent's style of movement, balance, and strategy.

This tactic is described as a maneuver in which you change the pace of your offense or defense in hopes of confusing your opponent as to your true intentions. It may be that you initiate an attack and, as your opponent reacts in a defensive manner, you stop your attack long enough for him to cease his response; you then violently continue your initial attack. This tactic is very confusing to the uninitiated adversary, and it can be used continually during an extended confrontation.

Break-rhythm fighting works exceptionally well with other fighting tactics presented in this chapter.

ERRATIC MOVEMENT

This fighting tactic can be employed against unarmed or armed adversaries at close- or mid-range distances. The purpose of this maneuver is to confuse your opponent and force him to misjudge his target.

In armed combat situations, the effectiveness of erratic movement is not 100 percent guaranteed because there are several uncontrollable factors. If you are moving toward or away from an armed opponent, his weapon expertise, aim, and other factors will determine your ability to evade his projectile. Regardless of your opponent's range and expertise, your erratic movement should be erratic even to yourself. This increases your chances for success and decreases the chances of your subconsciously setting up a predictable pattern of movement.

It may be necessary to crawl, drop, spin, or move from side to side, diagonally or horizontally. The more erratic and unpredictable your movement, the better your chances are of avoiding being stuck by a projectile.

Erratic movement will work well with other close- and mid-range armed or unarmed fighting tactics found in this chapter.

TIRED SYNDROME

The tired-syndrome tactic is best employed against unarmed, close-range adversaries. However, it can be used rather effectively against the armed opponent, provided the range and situation are appropriate for its use.

In unarmed confrontations, this tactic is used to make your opponent think you are completely exhausted, out of breath, weak, or lacking endurance. In many cases, this will cause your opponent to move in for the kill to take advantage of your diminished physical state. As your opponent moves in with an overconfident attitude, you spring into action with an accurate and precisely delivered technique.

The unarmed tactician can usually apply this method if the confrontation has been going on for an extended period of time. One should realize the importance of not overusing this method against the same opponent.

This method will work well in conjunction with other forms of close-range, unarmed tactics.

INJURIUS FAKUS

Injurius fakus is a term I coined to describe the following fighting tactic. Essentially, it involves faking an injury with the intent to force the armed or unarmed adversary to move in for the kill. This method works more effectively in close-range, unarmed encounters because there is less risk involved than in armed encounters.

Most adversaries, out of survival instinct, will feel the urge to close in and finish off the encounter when they see a weakened opponent, and the effectiveness of injurius fakus relies upon this instinct. If maximum effectiveness is to be derived from this method, the opponent must have reason to assume you have been struck by him or his weapon. As he comes closer to investigate or continue his aggressive assault, remain alert and counter when the opportunity presents itself.

Accurately and precisely executed countermeasures must be swiftly taken if success is to be achieved from this tactic.

Injurius fakus will work well in conjunction with other unarmed and armed close-range encounters.

VARIANCE

The variance form of fighting has been mentioned

because it is necessary for the shuriken artist to be aware of Eastern martial art tactics, as well as Western fighting methods.

Any of the tactics described in this book could be mastered and utilized in a specific manner to achieve desired results, but the well-rounded fighting tactician should be well versed in many strategies. He must be able to apply them in unpredictable combinations against an opponent. This in turn would ensure that his game plan would never be discovered by even the most skilled fighter, regardless of whether or not his opponent is armed.

Learn to use the variance methods, and you will be successful against a wide array of diversified opponents.

MULTIPLE OPPONENTS

Since there are so many possible situations which arise in armed or unarmed confrontations, it is not the purpose of this book to offer all possible solutions, but rather to make the shuriken artist aware of several basic strategies by which most problems can be solved.

Many tactics covered here can apply to one or more opponents in the same confrontation. When dealing with multiple opponents, your method and direction of movement will be of great importance, and in many cases may determine whether you come out of the situation unscathed.

Always move so that one or more of your opponents is constantly getting in the other's way, thereby blocking his partners' direct line of sight. Circular movement is the most effective method for multiple-assault situations because it makes your opponents crowd their partners. Never let two opponents be equally distanced from you, especially in close-range encounters. You should always attempt to eliminate the opponent positioned nearest to you, and keep the downed opponent between you and your remaining adversaries. All tactical ploys should be spontaneous and unpredictable; don't worry if all ploys do not work in the manner you originally intended.

TACTICAL DO'S AND DON'TS

Become familiar with the following "do's" of fighting:

DO

1. Watch your opponent at all times.
2. Remain sure-footed and on solid ground as much as possible.
3. Use your peripheral vision to look for chances to use objects tactics.
4. Use the fighting tactics in this book.
5. Use distraction methods against your opponent.
6. Develop systematic rapid-fire techniques.
7. Use martial arts training in addition to shuriken tactics when your opponent is in close range.
8. Use the terrain to your advantage.
9. Use the sun to your advantage whenever possible.
10. Be aware of the range between you and your opponent.
11. Use fakes and half-commitments to force your opponent to react.
12. Fight dirty.
13. Emphasize penetration techniques in combat situations.
14. Keep vital areas well protected.
15. Use only techniques that you have mastered thoroughly.

The following "don'ts" could very well save your life:

DON'T

1. Use fancy maneuvers or techniques.
2. Let the shuriken stay in one position for too long.
3. React hastily.
4. Take your eyes off your opponent.
5. Remain in one posture for a long period of time.

Fighting Tactics

6. Use fancy or complicated footwork.
7. Throw your last shuriken at an opponent when he is still armed.
8. Get forced or trapped in a position in which you cannot readily move about.
9. Waste effort on impractical situations.
10. Let your guard down while changing positions.
11. Make full-commitment movements when your opponent is aware of your tactic.
12. Aim for targets which will have minimal effect.
13. Be overconfident.
14. Neglect your opponent's range and skills.
15. Forget to use stealth tactics whenever possible.

15. Target Areas

You can readily strike hundreds of target areas on the human body. These areas are classified into three groups: nonlethal, semilethal, and lethal areas. Depending on the severity and depth of penetration, however, many attacks to lethal or semilethal areas could possibly become lethal if immediate medical attention is not administered.

From a combat point of view, even though the standard star- or spike-shaped shuriken is not considered a lethal weapon by design, the depth or location of a puncture created by these "nuisance" weapons can cause a wound to be lethal.

It should also be noted that in times past, the Ninja certainly must have been aware that shuriken blades, when dipped in various poisons, become all the more lethal.

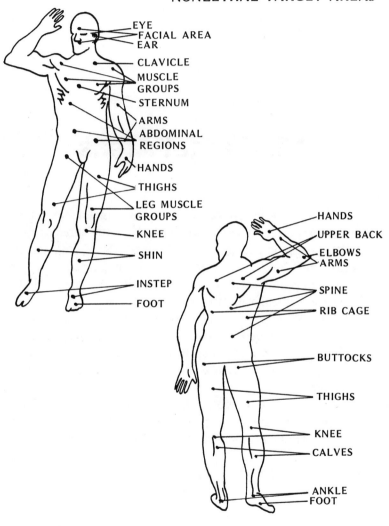

NONLETHAL TARGET AREAS

Severe pain and the inability to function responsively occur when one is struck in a nonlethal target area. Such injury is normally inflicted by shurikens that do not possess deep penetrating effects.

SEMILETHAL TARGET AREAS

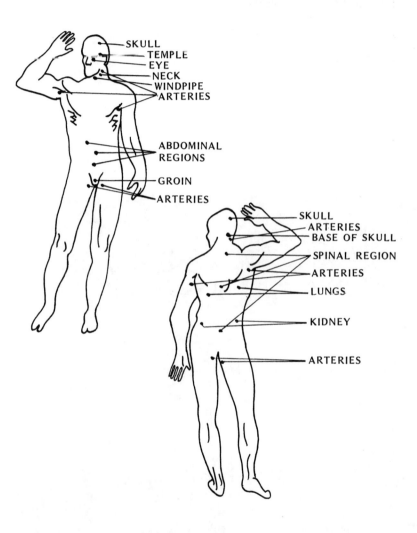

A forceful strike of moderate depth to the above semilethal target areas with a star or spiked shuriken can be lethal if urgent medical attention is not received.

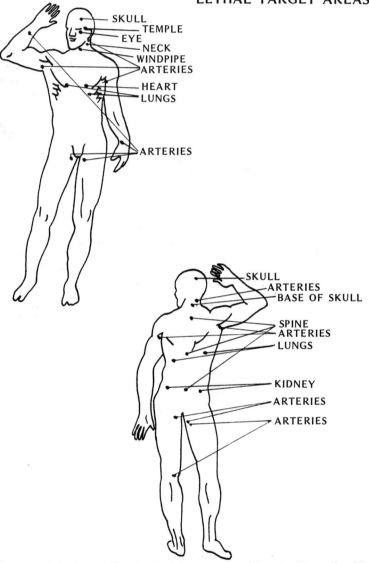

LETHAL TARGET AREAS

Deep, penetrating strikes to the above areas with a shuriken of sufficient length can be lethal if immediate medical attention is not received.

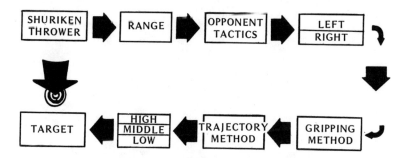

THE FLOW CHART
The above flow chart indicates the processes that must be understood and mastered before one can use the shuriken.

16. Shuriken Concealment

The star or spike shuriken can be concealed in many ways. It should be remembered, however, that a weapon should be concealed in such a way that it is readily accessible when required. Otherwise, the purpose of having a weapon is thoroughly defeated.

Several custom-designed shuriken presently on the market can be folded or collapsed to fit conveniently into wallets, watch pockets, or other small spaces in garments or accessories.

The skilled shuriken artist will conceal star and spiked shuriken in different locations, which he can put into play regardless of his position.

Special leather pouches can be designed to carry several star-shaped shuriken in much the same fashion as a wallet. These can be brought into action at a moment's notice. Spike shuriken can also be disguised as pencils or pens and carried inconspicuously and in open sight without being detected. Many shuriken artists prefer to have special pouches that attach to the ankle, inside the shirt sleeve, or in other readily accessible locations.

Shuriken have been classified as deadly weapons in several states. In these states, possession is illegal, and there is an additional penalty for weapon concealment. Possession

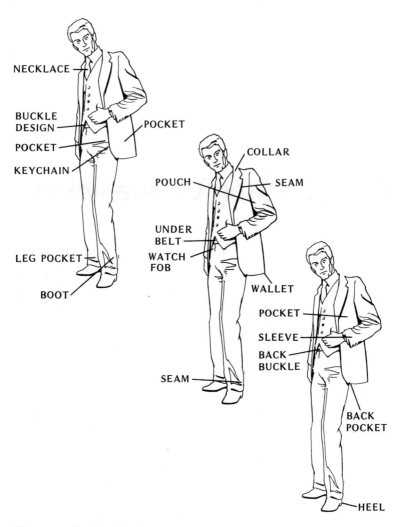

The star and spike shuriken can be concealed in various locations, some of which are illustrated above.

and concealment of a deadly weapon is a felony, and arrest on this charge would create problems for a citizen wishing to maintain a clean record.

17. Moving Targets

Accurately penetrating a moving target is perhaps the most difficult of all objectives when throwing the shuriken, partially because there are many factors which are beyond your control. By understanding these factors it may be possible to select the type of shuriken which will increase the possibility of success in such situations.

First, one must realize a moving target can either be moving toward you or away from you. In the event that it is a suspended target, you will be confronted with a certain amount of spin in addition to the aforementioned factors. Unless you have developed an acute sense of timing, it is very unlikely you will penetrate the target when it's in motion.

When the target is moving away from your position, the velocity of the shuriken will diminish greatly after it has exceeded the range of preselected trajectory. Even the most sophisticated of multipointed shuriken will then fail miserably at its task. The closer the moving target, the greater will be your chances of making a successful strike.

If the surface of the target is made of wood, the slightest tilt, forward or backward motion, or spin will lessen the possibility that the multipointed shuriken will penetrate the target surface. To overcome most of this difficulty with the multipointed shuriken, the velocity or amount of power emphasized in the throw must be increased more than otherwise required should the target remain completely at rest. If you are using the single-point spike, this method of practice will be extremely difficult. Using the double-pointed spike, the odds still favor the target. With spike shuriken, factors such as the amount of spin, distancing, and velocity must be mastered before consistent accuracy is achieved.

In the event you select a multipointed star shuriken, you must realize that the increase in points does not necessarily make it less difficult to strike a moving target. If there are many points which are positioned very closely together, increased velocity in the throw will generally increase the amount of spin on the shuriken, thereby causing the weapon to continue spinning at the time of impact. The points that subsequently strike the target will only cause the shuriken to pull itself out of the target after it has made penetration.

This information should always remain at the forefront of your thoughts when you select a shuriken for use on a moving target. Over the years, I have learned to use a rather heavy, solid form of shuriken with five equidistant points.

It is wise to be aware of the necessity of additional training when throwing a shuriken at a moving target in order to get accustomed to anticipating the target's movement. This form of training will increase your possibility of success. With free-form moving targets, such as a person or small game, any empty-hand martial arts training will help you to become aware of the target's intentions.

Another important but seemingly insignificant factor in becoming adept at hitting a moving target is to become thoroughly conscious of the amount of time actually required for the shuriken to leave your hands and impact with the target. Even if this time is in milliseconds, it will give you some reference to the amount of power and velocity necessary to achieve maximum penetration even under adverse conditions (such as movement and spin).

Second, consistency is achieved by standardizing the type

of shuriken that is used against a moving target. If you have a broad selection of shuriken with different weight characteristics, you will have some idea as to how much power and velocity is needed to minimize the amount of time it takes the weapon to reach its target.

One of the unique aspects of the shuriken is that even under the adverse conditions, your chances are generally better than if you were using a knife or other single-bladed throwing device. This aspect, coupled with the fact that the experienced shuriken thrower will have a selection of shuriken, gives the added advantage of more than one chance to hit the target.

MOVING TARGET CRITICAL FACTORS

1. TARGET MOVEMENT
   A. Toward you
   B. Away from you
   C. Stationary at initial point of movement
   D. Diagonal forward motion (left or right)
   E. Diagonal backward motion (left or right)
   F. Upward (jumping motion)
   G. Downward (dropping motion)
   H. Turning or spinning maneuvers
   I. Deceptive (combination of two or more of the above)

2. DIMINISHED VELOCITY
   A. As target moves away, shuriken velocity diminishes.
   B. As target moves away, more power is needed in the throw.
   C. As target moves in, penetrating effects increase.

3. SURFACE TEXTURE
   A. The more dense the surface texture, the more difficult the shuriken penetration.

B. The less dense the surface texture, the easier the shuriken penetration.
  C. In wooden targets, more effective penetration is achieved if the shuriken is struck parallel to the grain.
  D. If target area is tilted, surface texture becomes more critical to the effectiveness of the shuriken.
  E. The more dense the surface area, the more power is required for adequate penetration.

4. WEAPON SELECTION
  A. Single- or double-pointed spike shuriken are more difficult to use than multipointed star types.
  B. Too much spin on multipointed shuriken will increase the possibility that the weapon will fail to penetrate the target.
  C. The weight of the shuriken becomes a critical factor.

5. PREJUDGMENTS
  A. Maximum effectiveness is achieved if prejudgment of the moving target is possible.
  B. Empty-hand martial arts training will increase awareness of the moving target's next movement.

6. TIME/DISTANCE RATIO
  A. Be conscious of the length of the shuriken's travel time to the target.
  B. Be aware of the weight of specific shuriken types to determine the amount of power needed for distance.
  C. Use accumulated experience to make proper judgments.

7. STANDARDIZED SHURIKEN
  A. To develop consistent accuracy, shuriken weight should remain the same.
  B. Consistency and accuracy is achieved through practice.

18. Rapid Fire

The rapid-fire method of shuriken trajectory is unique in that the shuriken can be continuously dispensed in the blink of an eye. The speed at which it is thrown is determined by several characteristics, both in the skill of the thrower and in the design of the weapon itself.

One should be aware, however, that because the science of using and throwing the shuriken is an exacting one, certain sacrifices must be made if rapid-fire dispensing is desired. The more speed gained in the dispensing, the more power will be lost, primarily due to the fact that most rapid-fire techniques require the throwing wrist and arm to be involved directly with the trajectory. Maximum power is derived by using the sequential muscular method of trajectory, but this takes more time to accomplish and diminishes the ability to deliver the shuriken in rapid succession.

If the shuriken are positioned on a pintle (a spike that is used to mount shuriken that have a hole through the center), one hand will be required to feed the shuriken upward while the other hand dispenses them. Reliable dexterity must be mastered before consistent accuracy can be achieved.

If one hand is used to hold several shuriken instead of the pintle spike, it will be necessary to place the shuriken so that they overlap each other. This is to ensure that each shuriken can be gripped individually by the hand that is performing the throwing maneuvers. The holding hand must be placed in a comfortable and convenient position so that maximum speed can be achieved with the throwing hand. Usually the closer the holding hand to the throwing hand, the better. Since rapid-fire trajectory methods are quicker than single-throwing techniques, a considerable amount of power will be thereby lost. These rapid-fire techniques are not normally effective at greater distances. By overexcelling in reaching greater distances, accuracy and consistency will be lost in most cases.

The points or blades of the shuriken should be positioned so that the quick-retrieval of the second and consecutive shuriken will not puncture or slice the throwing hand. This is the worst aspect of rapid-fire throwing, and calls for extreme caution.

19. Multiple-Shuriken Throwing Techniques

The gripping methods presented in this book will be of great help in determining how to select a method that suits the shuriken practitioner's needs when throwing two or more shuriken at the same time. The style, weight, range of target area, gripping method, and pressure of grip are all factors that must be thoroughly understood before consistency can be achieved in this method of throwing. The shuriken artists who are most successful at this throwing method are the ones who have spent countless hundreds of hours and a great deal of trial and error in discovering which methods work most effectively for them.

Because of the way the shuriken is designed, when multiple-shuriken throwing is practiced, the greater the range, the broader the travel pattern of the weapons. This is similar to a shotgun: the greater the range, the wider the pattern of the shot.

In the shuriken, this characteristic is primarily due to the firmness with which they are held. If they are held quite firmly, the pattern will remain smaller for a bit longer, and if they are held loosely, the shuriken will tend to spread at a much faster rate. The thinner styles of shuriken have a tendency to separate at a much faster rate than the heavier, thicker ones. Until the patterns and separation factors of a

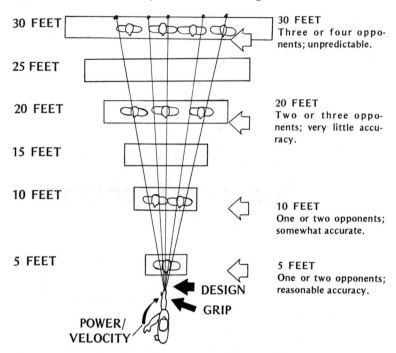

Design will determine the rate by which the shuriken separate (the lighter or thinner the weapon, the more it will separate). Power and velocity also affect the separation rate of the shuriken (the more power velocity applied, the less immediate separation).

given type of shuriken are thoroughly understood, it will be very difficult to accurately predict its flight and trajectory.

It should also be noted at this point that the more shuriken thrown at one time, the less control and predictability can be expected from the weapons. Through years and many hundreds of hours, it has been my experience that three thrown at one time seems to be the maximum number that can be controlled with predictable accuracy. In most cases this has been at relatively close range.

The chart accompanying this section will be of importance in creating a clear pattern of the multiple-shuriken throwing characteristics and the variables that affect their trajectory.

20. Target Competition

Target competition is essentially a method of practicing the shuriken techniques presented in this book. You may want to consider them games of a sort, but the primary objective will be to help develop efficient skills with your weapon.

The competitive aspects of training with the shuriken can be very rewarding when you begin to develop proficiency in areas such as accuracy, spontaneity, rapid fire, consistency, trajectory techniques, variations in gripping methods, and a host of other qualities that are necessary before the art is truly mastered.

The competitive games covered in this section represent but a few of the possibilities. With a little imagination, one can devise others, or variations of the ones covered here. You should always have an objective or purpose for the type of competition that is devised, such as developing one or more of the qualities necessary to increase your overall proficiency with the weapon. Never play just for the sake of fun.

POINTS

A target should be constructed consisting of a variety of target areas with ratings of different point values. The higher values should be near the center of the target.

Two or more players, each using three shuriken, will throw their shuriken in turn, and then tally the points scored in each round. A designated number of rounds should be set before the competition begins, and distances from the target can vary.

If the higher point values are toward the center of the target, one of the primary purposes of this type of competition is to develop consistent accuracy at a designated target. If different ranges are established, it will give the throwers a chance to develop variances in range, which is a very important aspect of moving target practice.

ROTATION

Specific areas of the target (either by numerical order or by location of a human anatomy target) are selected, and players alternate in attempting to hit different areas. The one with three "lethal" throws wins the round.

ATTACKER

From any of the stances or throwing positions, the players draw and throw the shuriken, all beginning at the same time (with someone to begin each round) and "draw and fire" the weapon in the shortest possible amount of time. This method can be scored in a variety of ways, but accuracy should be an important consideration if benefits other than speed in delivery are to be developed.

SURPRISE ATTACK

This practice method consists of players alternately standing at a specific distance with their backs to the target. On command, the player should turn as quickly as possible and throw at the target. There are many variations of this theme, but the primary objective is to develop reflexive skills and shorten reaction time.

MULTITHROW

This method employs two or more shuriken which are to be thrown at the same time. With practice, distinct characteristics utilizing different gripping methods will be revealed. If the human anatomy type of target is used, points can be accumulated on the number of target areas struck with the shuriken in each round of competition.

BLINDFOLD PRACTICE

This method of competitive practice is unique in that players develop a strong understanding of concentration and a real feel for the shuriken. Each player in turn, while blindfolded, positions himself in a location relative to the target, visualizes the target area, and throws at specific target areas. The player who accumulates the most accurate strikes is the winner. There can be as many rounds as have been designated before the competition begins. Variations in range and positioning can be very efficient ways of practicing this method.

IMITATION

A player is selected to begin. This player throws his shuriken at any position on the target that he desires; then the following player tries to strike as close to the first player's throw as possible. The one who achieves the most close position impacts after a predetermined number of rounds is the winner. This method of practice is useful in developing instant target recognition and variations in target positioning. Many variations of this theme can be played.

21. Additional Shuriken Considerations

With the popularity of the shuriken reemerging in the twentieth century, it is quite probable that many new designs and innovative methods of propelling them will be developed.

Many modern-day concepts concerning shuriken designs are already beginning to surface in certain publications, and creative methods of unique design are being custom-made by prominent knife makers.

Perhaps the most innovative use of the shuriken is in the collapsible styles that can be folded open in much the same manner as a pocket knife. The blades or points of the four-point variety are designed in such a way that a spring-loaded bolt is positioned through the middle of the blade so that the blades can be locked in place in the open position, thrown, and then, with a bit of finger pressure at the connecting bolt in the middle of the shuriken, collapsed and folded into a small, concealable, double-pointed, flat-bladed instrument.

These innovations, and especially the style of shuriken mentioned above, are very popular with many who have had prior martial arts experience. These weapons are unique in the fact that they can be easily concealed and brought into play in the blink of an eye. When folded, these shuriken are usually about three inches long. They are made of high-quality steel and have extremely sharp blade and point sections.

Similar in design, the longer version (about eight inches), is carried by some on combat missions. These are perhaps the most lethal of all shuriken types existing today, for they can penetrate at a depth of up to four inches. They also possess ample weight to make that penetration possible, requiring very little power or velocity to propel the weapon.

Mercenaries have rediscovered the importance of the shuriken as a silent and effective way of dealing with an enemy in close- or short-range combat situations. Most varieties are painted black so they are nonreflective.

There will perhaps come a time when many ancient styles of weapons will be propelled by handheld devices designed in much the same manner as a pistol. Quite possibly they may be used with a type of powder bullet which serves as the source of their propulsion.

I have had some rather unique experiences in using a slingshot as a method of propelling the shuriken, and have realized that with a sufficiently weighted shuriken it is possible to achieve uncanny accuracy and a variety of curving characteristics. There is no doubt that an incredible amount of penetration can be developed in throwing the shuriken in this manner.

It is also possible that a projector similar to a crossbow could be used to generate an even greater amount of velocity and penetration. It is also quite possible that projectors could be designed that would fire the shuriken in a rapid-fire manner similar to a rifle. By using carbon dioxide or other gasses, these would not only be relatively silent, but quite accurate at a great range as well.

These styles would certainly take the shuriken out of the realm of "nuisance weaponry" and put them in the category of *lethal weapons.*

Other interesting shuriken concepts are the use of the shuriken in belts, belt buckles, and as a decorative ornament or necklace design. These seemingly novel concepts can be utilized effectively provided that the shuriken have a practical design.

22. Conclusion

In concluding this text, I would like to mention that it has been my intention to make the reader aware of the many diverse facets involved in martial arts study.

The word *art* is in essence a misnomer; *science* is perhaps a more accurate term to describe the exactness and thoroughness necessary for one to become deeply knowledgeable in these studies.

To the layperson, the shuriken looks like a simple toy, yet the individual familiar with its use will recognize it as a deadly weapon when in the hands of a skilled practitioner. As with all martial arts forms, much practice, constant evaluation, and perseverance are required to gain expertise with the shuriken.

Know yourself and your limitations, and then attempt to overcome any of the shortcomings you may have. These efforts will not go unrewarded if every conscious effort is made to reach beyond your present capabilities, and you strive for a greater goal in both mental and physical realms. I hope the task of mastering the techniques and methods presented in this book will make you aware of what can be accomplished if the desire is great enough.

It should always be remembered that books are primarily a supplement for training, and they should never be considered the ultimate way of learning any of the martial arts.

I wish you the best of luck and success in any of the martial arts you seek to master.